YOUR KNOWLEDGE HAS VALUE

- We will publish your bachelor's and master's thesis, essays and papers

- Your own eBook and book - sold worldwide in all relevant shops

- Earn money with each sale

Upload your text at www.GRIN.com and publish for free

Bibliographic information published by the German National Library:

The German National Library lists this publication in the National Bibliography;
detailed bibliographic data are available on the Internet at http://dnb.dnb.de .

Imprint:

Copyright © 2018 GRIN Verlag
Print and binding: Books on Demand GmbH, Norderstedt Germany
ISBN: 9783668876897

This book at GRIN:

https://www.grin.com/document/453957

Hans Georg Schrey

Effect of Fouling on Performance of Exchangers Cooled by Air. Ramifications for Exchanged Heat and Cooling Effectiveness

GRIN Verlag

Effect of Fouling on Performance of Air-Cooled Exchangers

Author Hans Georg Schrey

On behalf of all authors, the corresponding author states that there is no conflict of interest.

There are no other authors.

Table of Contents

Abstract 3

1. Introduction 4

2. General Fin Tube Geometry 6

3. Heat Balance 12

4. Pumping Power 14

    4.1   Bundle Pressure Drop 15

    4.2   Natural Draft Pressure Gain 15

    4.3   Air Inlet Pressure Drop 15

    4.4   Air Outlet Pressure Drop 16

5. Determination of Air Flow 16

6. Results Discussion 17

Glossary 24

Bibliography 28

<u>Abstract</u>

During operation of heat exchangers layers of deposits or corrosive products may be formed and accumulated on heat exchanger surfaces over time. This leads to additional heat transfer resistance and constriction of fluid flow area. In consequence, the exchanged heat duty is badly affected. The loss of heat duty is extreme if local heat transfer coefficients are high at clean conditions. However, maintaining cooling effectiveness is paramount in most applications. As a remedy, surfaces must be regularly cleaned. The frequency of cleaning depends on rate and type of fouling. Depending on the type of heat exchanger various procedures and devices for cleaning surfaces have been developed over the past years.

Fin tubes are core elements in air cooled exchangers or condensers to transfer heat. Fin tube exchangers are characterized by a multitude of circular, elliptical or channel type core tubes with air-side finning. Generally, the process medium flows on the tube internal side with air as coolant on the external fin side. The report deals with air cooled heat exchangers and condensers under forced or natural draft in dry cooling applications with the focus on the effect of fin side fouling. Water spray injection into the cooling air flow is excluded.

The effect of fin side fouling layers will be assessed as well as the consequence for air flowrate and heat duty at different convection types. Special attention is given to the effect of fouling on the performance of dry air-cooled condensers. Also, differences of forced, induced or natural draft dry cooling applications will be covered.

<u>Keywords</u>:   Scaling; Fouling; Fin Tubes; Air Cooled Condenser; ACC; Air Cooler; Natural Draft; Vacuum Decay; Heat Duty; Mixed Convection

1. **Introduction**

Fouling has a detrimental effect on heat exchanger performance. In case of dry air-cooled exchangers specifically two factors are relevant on the fin side:

(a) Increase of heat transfer resistance by formation of a fouling layer on surfaces,

(b) Constriction of air side flow area, loss of cooling air flow.

Generally, fin side heat transfer coefficients are in the range of only 35 to 50 W/m²K whereas tube side coefficients may be 20 to 100 times higher. Large surface extensions are needed to make up for this discrepancy. Compact finning with small inter-fin gaps (in the low mm range) are a typical consequence of surface enlargement.

As for heat transfer resistance - fin side fouling resistance of typically 1/2000 m²K/W or less has only marginal influence on total heat transfer (fig. 1). For typical design values of NTU and fouling it is easy to see that the effect of fouling layer heat transfer resistance alone accounts for much less than 1% over the typical range of fin side heat transfer coefficients and fin surface extension ratio, φ. This is in contrast to the tube side where typically much higher heat transfer coefficients will be severely affected by tube side fouling.

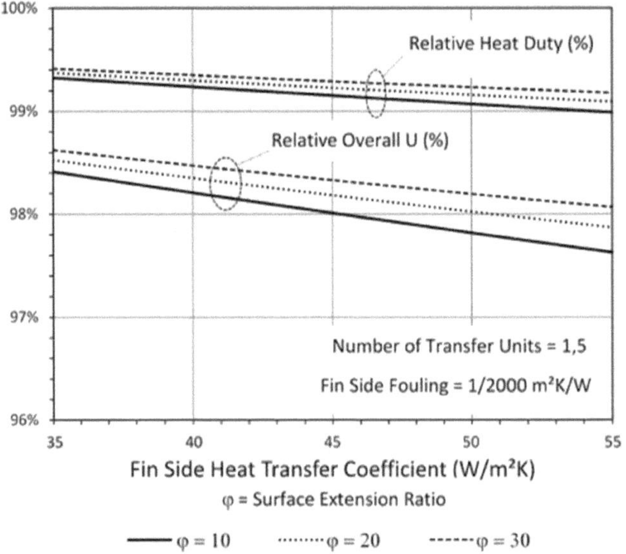

Figure 1: Thermal Effect Caused by Fin Side Fouling (source: own diagram)

Therefore, for simplification we shall neglect factor (a) and focus on factor (b) in the following considerations.

Constriction of flow area leads to loss of cooling air flow caused by increase of air side flow resistance. This may be the result of two factors:

(1) Blocking of air side gap area by pollen, leaves or other botanical matter resulting in total loss of cooling air flow and active fin surface locally – inhomogeneous fouling

(2) Reduction of inter-fin cooling passages by formation of layer of deposits on fin surfaces – homogeneous fouling (scaling)

Both effects lead to air speed acceleration in the residual non-blocked flow areas and increase of air side pressure drop. As a result, the air outlet temperature increases and the projected heat duty and process exit temperature can no longer be reached.

In case of air-cooled condensers (ACC's) the un-wanted loss of cooling air flow goes along with increase of vacuum pressure and reduction of generated power – i.e. loss of profit. Therefore, plant operators ensure performance by regular cleaning of fin side surfaces. However, expenses for cleaning must be weighed against the anticipated advantages. So, there is a need to understand the probable loss of profit resulting from loss of performance. The general inter-dependency of effective heat rejection and air flow rate must be known.

In the following, for simplification we shall take - wherever possible - all operation parameters at design conditions to identify the net fouling effect. Only steady state conditions will be considered. Furthermore, in case of forced or induced draft all fans shall be running at design speed. If not stated otherwise general model and simplifications will be assumed in line with well known standards, such as [1] or [2]. Quotation [3] shows a detailed derivation for the case of forced draft air-cooled condensers (ACCs). In addition, air side temperature rise will be incorporated to estimate the natural draft effect. The proposed calculation procedure is intended to be workable under a great variety of fin tube arrangements. It covers cross-counterflow arrangements with single or multiple tube rows with core tube geometries of circular, elliptical or channel shape. There is also no restriction to selected fin geometry type.

2. <u>General Fin Tube Geometry</u>

Fin tube systems cover a wide variety of core tube and finning geometries. All these combinations have one thing in common: the minimum inter-fin tube gap area constituting the bottleneck for airflow (fig. 2). Apparently, the fin boundary layer formed at this location predominates the thermo-hydraulic characteristics of the whole system. Therefore, the dominating Reynolds number of fin tube systems is defined at this minimum inter-fin gap by the bulk of proposed fin tube correlations.

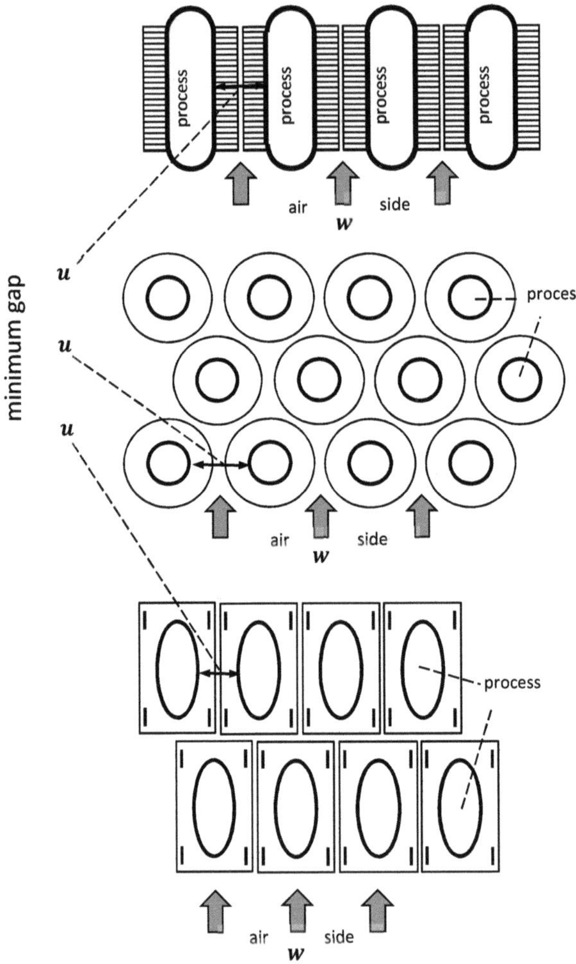

<u>Figure 2: Typical Fin Tube Systems (source: own sketch)</u>

The following parameters are common to all types of fin tube systems (fig. 3):

- minimum inter-fin tube gap area, A_g

- frontal air side face area, A

- finned surface, F_a

- internal tube side surface, F_i

- transverse maximum external core tube diameter, d_t

- transverse maximum external over-fin diameter, D

- transverse tube pitch, p_t

- fin pitch, p_f

- fin thickness, s_f

- fin root thickness, δ_f

- coating thickness, δ_c

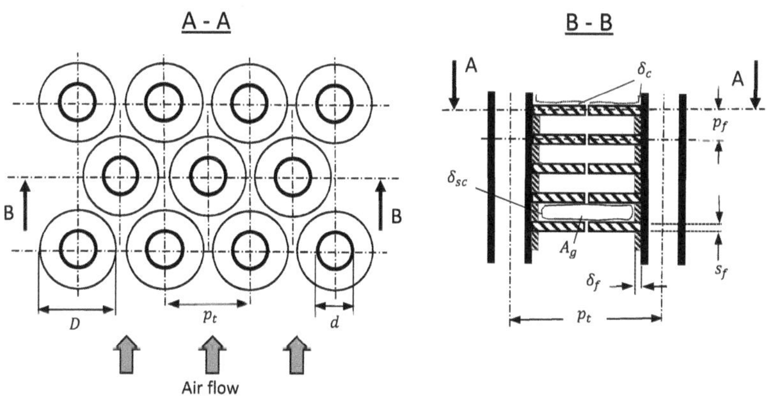

Figure 3: Fin Tube Geometry (source: own sketch)

Although figure 3 shows only the example of a standard circular fin tube with circular finning the parameters may easily be adopted for other geometries – elliptical or rectangular. In case of fouling there is an additional

- external scaling layer, δ_{sc}.

The geometry above covers all well-known combinations of fin and core tube connections, such as

- fin sheet or lamella fin
- integral fins (I fin)
- grooved fins (G fin)
- bended fin foot (L fin), overlapped finning
- extruded fins
- galvanized fins, coated fins

To account for these variations some parameters must be adjusted. Table 1 shows the parameter variations in question (cf. fig. 3, "B-B"). Considering these adjustments we can formally restrict the general geometry to "I" fin parameters plus extensions.

Clean Fin Tubes **Fouled, Scaled Fin Tubes** [*]	I fin G fin	L fin Single Row	Extruded fin	Coated fin
Fin root thickness	$\delta_f = 0$	$\delta_f = s_f$	$\delta_f = s_{ext}$	$\delta_f = \delta_c$
Open inter-fin pitch	$p_{f0} = p_f - s_f$			$p_{f0} = p_f - s_f - 2\,\delta_f$
Over fin tip gap	$s_{tip} = p_t - D$			
Fin root thickness [*]	$\delta_f^* = \delta_f + \delta_{sc}$			
Open inter-fin pitch [*]	$p_{f0}^* = p_f - s_f - 2\,\delta_{sc}$			$p_{f0}^* = p_f - s_f - 2\,\delta_f^*$
General		Clean		Fouled [*]
Maximum transverse core tube diameter		$d = d_t + 2\,\delta_f$		$d^* = d_t + 2\,\delta_f^*$
Open inter-tube height		$h_{t0} = p_t - d$		$h_{t0}^* = p_t - d^*$

<u>Table 1: Geometric Effect of Fin Root Type (source: own sketch)</u>

In case of fin sheets or lamellas there is no over-fin gap so that $D_{(sheet)} = p_t$.

Fouling shall form only on the flat surfaces of the fin tubes. If the fin tip gap is filled with matter a small effect on face area ratio shall be considered in the following.

At this point we have to make simplifying assumptions concerning the geometric effect of scaling layers. First of all, the scaling layer constitutes additional air flow resistance. Secondly, there is an effect on active heat transfer surface. As long as the scaling layer is thin the effect may be neglected. Roughness of the scaling makes up to some extent for the loss of plain surface. On the other hand, as fig. (3, "B-B") suggests we need some allowance for loss of heat transfer surface if a scaling layer develops. In view of the many geometries involved as well as the multitude of possible fouling distributions in a tube bundle a simplification may be justified. We assume proportionality between contact circumference in the flow gap and active heat transfer surface.

$$\frac{F_a^*}{F_a} \sim \frac{p_{f0}^* + (h_{to}^* - s_{tip})}{p_{f0} + (h_{to} - s_{tip})} = f \tag{2-1}$$

With this assumption the surface loss effect is over-estimated to some extent. On the other hand as we have disregarded the additional heat transfer resistance by the fouling layer the simplification may be justified.

Special consideration must be taken in case of partially blocked airflow. For better understanding see fig. (4) showing a typical cross-counterflow arrangement. Contrary to process flowrate only part of the air side is active, namely

$$\beta = \frac{A_b}{A} = \frac{A_{gb}}{A_g} = \frac{A_{gb}^*}{A_g^*}$$

This pertains to air flow as well as total exchanger surface. On the other hand, area ratios remain unaffected by blocking (cf. fig. 3, B-B).

Face area ratio

$$\Omega = \frac{A}{A_g} = \frac{A_b}{A_{gb}} = \frac{p_f \cdot p_t}{p_{f0} \cdot h_{to} + s_{tip} \cdot s_f}$$

Surface density

$$\sigma_a = \frac{F_a}{A} = \frac{F_{ab}}{A_b}$$

If a scaling layer is formed surface density changes to

$$\sigma_a^* = \frac{F_a^*}{A} = f \cdot \sigma_a$$

and face area

$$\Omega^* = \frac{A}{A_{gb}^*} = \frac{A}{\beta \, A_g^*}$$

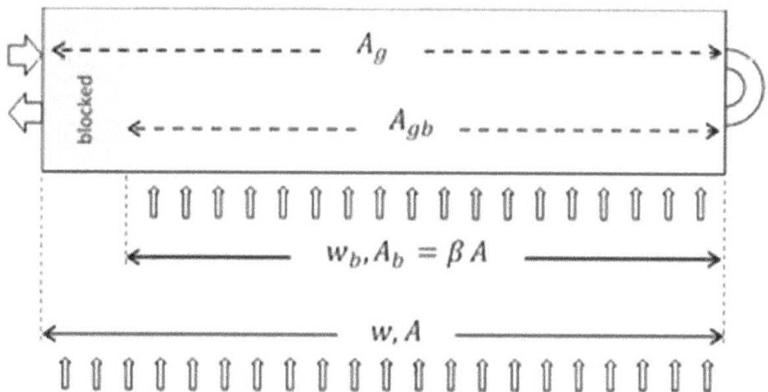

Figure 4: Blocked Bundle Geometry (source: own sketch)

The relative flow area restriction caused by fouling is

$$\omega \equiv \frac{\Omega}{\Omega^*} = \frac{\beta\, A_g^*}{A_g} = \beta \cdot \frac{p_{fo}^* \cdot h_{to}^* + s_{tip} \cdot s_f \cdot (1 + 2\,\varepsilon)}{p_{fo} \cdot h_{to} + s_{tip} \cdot s_f} \qquad (2\text{-}2)$$

with the relative fouling layer thickness defined as $\varepsilon = \delta_{sc}/s_f$.

The air mass flow balance yields

$$\rho_1\, w\, A = \rho_1\, w_b\, A_b = \rho_m\, u\, A_{gb} = \rho_m\, u\, \beta\, A_g$$

with air inlet velocity w_b already being accelerated. Fouling affects total air and gap flow:

$$\rho_1\, w^*A = \rho_1\, w_b^*\, A_b^* = \rho_m^*\, u^*\, A_{gb}^* = \rho_m^*\, u^*\beta\, A_g^*$$

In clean operation the total face area is active. The total air mass flow ratio of fouling to clean is

$$\alpha = \frac{\rho_1\, w_b^*\, A_b}{\rho_1\, w\, A} = \frac{\rho_m^*\, u^*\beta\, A_g^*}{\rho_m\, u\, \beta\, A_g} = \frac{\rho_m^*\, u^*}{\rho_m\, u} \cdot \omega/\beta \qquad (2\text{-}3)$$

The capacity flow ratio – assuming approximately constant specific heat capacities of either side – may be expressed by a mass flow ratio. While air mass flow is reduced still, the full design process mass flow takes part in heat transfer (fig. 4). Consequently, only air flow changes the effective ratio.

$$\frac{V^*}{V} = \left(\frac{\rho_1\, w_b^*\, A_b}{c_P\, M_P}\right) : \left(\frac{\rho_1 w\, A}{c_P\, M_P}\right) = \frac{\rho_1\, w_b^*\, A_b}{\rho_1\, w\, A} = \alpha$$

Fin tube performance is commonly expressed in form of Reynolds number correlations which require the definition of a hydraulic diameter for dimensionless notation. Geometry effects are expressed by appropriate proportionality factors and exponents. Assuming that fouling does not change correlation constants and exponents the effective operation is restricted to Reynolds variation alone. Most important, the hydraulic diameter of the fin tube system need not to be defined. This is explained in detail in quotation [4]. As usual all air side properties will be taken constant at ambient air temperature – except for density which is calculated at mean air temperature. This leaves only gap mass flux $(\rho\, u)$ as operational variable.

Summarizing, the Reynolds number is $Re = \frac{D_h}{\eta_a} \cdot \rho\, u \sim \rho\, u$ or, at fouling (cf. 2-2, 2-3)

$$\frac{Re^*}{Re} = \frac{\rho^*\, u^*}{\rho\, u} = \alpha\beta/\omega \qquad (2\text{-}4)$$

Note that the type of fouling – either face area blocking or homogeneous flow passage restriction – lead to different total heat duty. With face area blocking the proportional part of the active heat transfer surface is blocked as well. For the purpose of comparison we shall consider both types separately (cf. table 2).

Fouling Variant (Fouling Distribution)	Blocking of Face Area (inhomogeneous)	Flow Passage Restriction (homogeneous)
Active Face Area, Exchanger Surface	$\beta < 100\%$	$\beta = 100\%$
Active Gap Area	$\omega = 100\%$	$\omega < 100\%$

<u>Table 2: Fouling Related Transfer Property Ratios (source: own sketch)</u>

To determine exchanger effectiveness the number of transfer units must be calculated. We have

(clean)
$$NTU = \frac{U \cdot \sigma_a}{c_P\, \rho\, u\, A_g}$$

(fouled)
$$NTU^* = \frac{U^* \cdot f \cdot \sigma_a}{c_P\, \rho^*\, u^*\, A_g^*}$$

So that
$$\frac{NTU^*}{NTU} = \left(\frac{U^*}{U}\right) \cdot \left(\frac{Re^*}{Re}\right)^{-1} \cdot \frac{A_g}{A_g^*} \cdot f \qquad (2\text{-}5)$$

By using (2-2) and (2-4) equation (2-5) yields the general NTU dependence.

$$\frac{NTU^*}{NTU} = \alpha^{m_k-1} \cdot \left(\frac{\beta}{\omega}\right)^{m_k} \cdot f \qquad (2\text{-}6)$$

Without sacrificing generality we have assumed that the air side dominates overall heat transfer: $U \sim Re^{m_k}$. The estimation of the overall U Reynolds exponent m_k is based upon fin tube heat transfer Reynolds exponent m. For details see §5.2 of [1]. Note that the recommendation $m_k = 0,45$ holds primarily for circular fin tube arrangements. In cases of single row systems this value is reduced to approximately $m_{k(SR)} \cong 0,3$.

At this point – as we need the air temperature increase to account for natural draft effects – we check the heat duty of clean and fouled system.

3. <u>Heat Balance</u>

Basically, we start with a typical cross-counterflow arrangement as common in air coolers and air-cooled condensers. The heat duty balance is

$$q = \frac{\dot{Q}^*}{\dot{Q}} = \frac{c_P\,\rho^*\,w^*\,A}{c_P\,\rho\,w\,A}\cdot\frac{\Phi^*\,\vartheta}{\Phi\,\vartheta} = \alpha\cdot\frac{\Delta T^*}{\Delta T} \tag{3-1}$$

Thus, the change of air side temperature difference may be estimated as

$$\delta \equiv \frac{\Delta T^*}{\Delta T} = \frac{\Phi^*}{\Phi} \tag{3-2}$$

The exchanger effectiveness of counterflow arrangements is a function of number of transfer units NTU and capacity flow ratio V, i.e.

$$\Phi = \frac{1-exp[(V-1)\cdot NTU]}{1-V\cdot exp[(V-1)\cdot NTU]} \tag{3-3}$$

In case of isothermal condensation ($V = 0$) this reduces to $\Phi_{is} = 1 - e^{-NTU}$.

Capacity flow ratio can be expressed as quotient of process and air side temperature difference. In standard applications air side temperature difference is larger than process side difference. We therefore restrict our consideration to cases where $V < 0{,}5$.

If we define $E \equiv e^{(V-1)\cdot NTU}$ then again, $E < 1$. Consequently, the quotient of the effectiveness function may be converted into the series $\frac{1}{1-V\,E} \cong 1 + V\,E + (V\,E)^2 + (V\,E)^3 + \cdots$.

If changes of (VE) are small the relative effectiveness at fouling is approximately

$$\frac{\Phi^*}{\Phi} \cong \frac{1-E^*}{1-E}\cdot\frac{1+V^*E^*+(V^*E^*)^2+(V^*E^*)^3+\cdots}{1+V\,E+(V\,E)^2+(V\,E)^3+\cdots} \cong \frac{1+(V^*-1)\cdot E^*}{1+(V-1)\cdot E}$$

or,

$$\frac{\Phi^*}{\Phi} \cong 1 + \frac{V}{e^{(1-V)\cdot NTU}-(1-V)}\cdot\left(\frac{V^*}{V}-1\right) - \frac{1-V}{e^{(1-V)\cdot NTU}-(1-V)}\cdot\left(\frac{E^*}{E}-1\right)$$

The exponent functions E can be linearized to

$$\frac{E^*}{E} = \frac{e^{(V^*-1)\cdot NTU^*}}{e^{(V-1)\cdot NTU}} \cong 1 - (1-V)\cdot NTU\cdot\left(\frac{NTU^*}{NTU}-1\right)$$

with the effectiveness ratio

$$\frac{\Phi^*}{\Phi} \cong 1 + \Gamma_V - \Gamma_N + \Gamma_N\cdot\frac{NTU^*}{NTU} \tag{3-4}$$

Here, we used the following definitions:

$$\Gamma_N \equiv \frac{(1-V)^2\cdot NTU}{e^{(1-V)\cdot NTU}-(1-V)} \tag{3-5}$$

$$\Gamma_V \equiv \frac{V\cdot(\alpha-1)}{e^{(1-V)\cdot NTU}-(1-V)} \tag{3-6}$$

In case of condensers ($V = 0$) auxiliary group Γ_V vanishes while Γ_N is equal to the definition of Γ in the literature, cf. [1] to [3]. Consequently, equations (3-4), (3-5) and (3-6) constitute a general effectiveness ratio of cross-counterflow arrangements. The published literature is only a special case of this general relation.

The change of temperature difference (3-2) is identical to the effectiveness ratio (3-4). Note that α plays only a role if $V > 0$. With change of NTU and the actual air flowrate α we can directly calculate the temperature difference.

In case of air-cooled condensers ($V = 0$) air temperature difference depends only on NTU. As an example, this was done in figure 5. We anticipate airflow acceleration within the minimum gap area which enlarges the local heat transfer coefficient and NTU. Consequently, exchanger effectiveness and air temperature difference rise as well. Loss of total air flowrate however makes up for this increase and heat duty goes down.

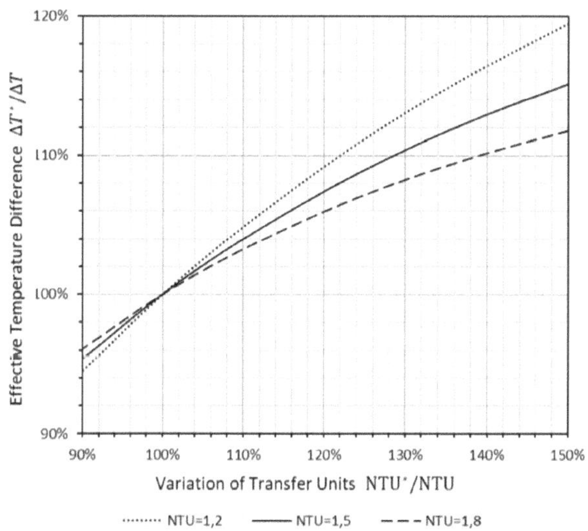

Figure 5: ACC Temperature Difference Effect of *NTU* Variation (source: own diagram)

Unfortunately, the graph cannot reveal the fouling effect on heat duty as the absolute loss of airflow is still unknown. To this end the airflow must be estimated by using the air side pressure drop balance as well as - in case of forced or induced draft - fan power balance. This will be dealt with in the next paragraph.

4. **Pumping Power**

Basically, fan pumping power is defined by fan volume flowrate and air pressure drop. As mentioned, inlet variables (such as ambient pressure or air inlet temperature) are kept at design conditions.

Irrespective of fan arrangement total air pressure drop remains the same. At small parameter changes the total fan power remains approximately constant

$$\dot{V}^* \cdot \Delta P^* \cong \dot{V} \cdot \Delta P$$

so that in forced draft:
$$1 \cong \alpha \cdot \frac{\Delta P^*}{\Delta P}$$

induced draft:
$$1 \cong \alpha \cdot \frac{\Delta P^*}{\Delta P} \cdot \frac{1 + \delta\,\tau}{1 + \tau} = \alpha \cdot \frac{\Delta P^*}{\Delta P} \cdot \theta_a$$

Definitions are
$$\delta \equiv \Delta T^* / \Delta T, \qquad \tau \equiv \Delta T / T_1, \qquad \theta_a \equiv (1 + \delta\,\tau)/(1 + \tau).$$

The pumping power equation may now be generally expressed as

$$1 \cong \alpha \cdot \frac{\Delta P^*}{\Delta P} \cdot \theta \tag{4-1}$$

where θ (table 3) is case dependent (forced or induced draft fans).

Correction Factor	*Forced Fan*	*Induced Fan*
Fan Arrangement	$\theta = 1$	$\theta = \theta_a \equiv (1 + \delta\,\tau)/(1 + \tau)$
Natural Draft Contribution	$\theta_b \equiv (1 + 0{,}5\,\tau)/(1 + 0{,}5\,\delta\,\tau)$	

Table 3: Air Density Corrections (source: own sketch)

Generally, airside pressure drop is a combination of friction losses and natural draft. As long as natural draft is small ($< 5\%$) as in most industrial applications the effect is not considered in well-known standards. For a change, we are going to include this effect in the following. Thus, the total pressure drop provided by fans is

$$\Delta P = \Delta P_B + \Delta P_1 + \Delta P_2 - \Delta P_N$$

If we define $\quad \Pi_B \equiv \Delta P_B^*/\Delta P_B,\ \Pi_N \equiv \Delta P_N^*/\Delta P_N,\ \Pi_1 \equiv \Delta P_1^*/\Delta P_1,\ \Pi_2 \equiv \Delta P_2^*/\Delta P_2$ and $\kappa \equiv \Delta P_N/\Delta P_B,\ \xi_1 \equiv \Delta P_1/\Delta P_B,\ \xi_2 \equiv \Delta P_2/\Delta P_B$ the pressure drop equation turns into

$$\frac{\Delta P^*}{\Delta P} = (\Pi_B + \xi_1\Pi_1 + \xi_2\Pi_2 - \kappa\,\Pi_N)/(1 + \xi_1 + \xi_2 - \kappa)$$

Natural draft factor κ comprises all combinations of draft: from pure forced flow ($\kappa = 0$) to full natural draft ($\kappa = 1 + \xi_1 + \xi_2$, no fan assistance). The general pumping power condition (4-1) takes the following form

$$(1 + \xi_1 + \xi_2 - \kappa) \cong \alpha \cdot \Theta \cdot (\Pi_B - \kappa\,\Pi_N + \xi_1\Pi_1 + \xi_2\Pi_2) \tag{4-2}$$

4.1 Bundle Pressure Drop

Fin tube pressure drop is governed by the air velocity in the smallest inter-fin gap. Using standard flow resistance correlations the bundle pressure drop is

$$\Delta P_B = \frac{C}{Re^n} \cdot \frac{\rho_m}{2} \cdot u^2 = \frac{C \cdot \eta_a^2}{2\,d_h^2} \cdot \frac{1}{\rho_m} \cdot Re^{2-n}$$

or,

$$\Pi_B = \frac{\rho_m^*}{\rho_m} \cdot \left(\frac{Re^*}{Re}\right)^{2-n} = \theta_b \cdot \left(\frac{\alpha}{\omega}\right)^{2-n} \tag{4-3}$$

with

$$\theta_b \equiv \frac{\rho_m^*}{\rho_m} = \frac{T_1 + 0,5\,\Delta T}{T_1 + 0,5\,\Delta T^*} = \frac{1 + 0,5\,\tau}{1 + 0,5\,\delta \cdot \tau}$$

The pressure Reynolds exponent of multi row circular fin tubes, elliptical fin tubes and single row designs ranges from $n \cong 0{,}33$ (circular fin tubes, [1]) to $n_{(SR)} \cong 0{,}7$ (single row).

4.2 Natural Draft Pressure Gain

The natural draft effect is a function of air temperature change.

$$\Delta P_N = (\rho_1 - \rho_2) \cdot g H_D = \rho_1 \cdot \left(1 - \frac{\rho_2}{\rho_1}\right) \cdot g H_D$$

As before, the density ratio may be expressed in form of temperature ratios

$$\Pi_N = \frac{1 - \rho_2^*/\rho_1}{1 - \rho_2/\rho_1} = \delta \cdot \frac{1 + \tau}{1 + \delta \tau} \cong \delta / \theta_a \tag{4-4}$$

4.3 Air Inlet Pressure Drop

Air inlet pressure drop covers all approach losses before entering the bundle. The value is not affected by fouling or blocking. As reference any flow area before the tube bundle may be selected.

$$\Delta P_1 = C_1 \cdot \frac{\rho_1}{2} \cdot \overline{w}_1^2$$

or, at fouling

$$\Pi_1 \equiv \frac{\Delta P_1^*}{\Delta P_1} = \frac{\rho_1^*}{\rho_1} \cdot \alpha^2 = \alpha^2 \tag{4-5}$$

A typical value for inlet loss fraction at design stage is $\xi_1 \approx 0{,}15$.

4.4 *Air Outlet Pressure Drop*

Air inlet pressure drop covers all exit losses after the bundle. Outlet losses are defined based on hot air exit velocity. The value is not affected by fouling or scaling – but blocking of exchanger area must be considered. As reference the effective tube bundle area must be taken.

$$\Delta P_2 = C_2 \cdot \frac{\rho_2}{2} \cdot w_2^2$$

The mass balance yields $\qquad \rho_1\, w\, A = \rho_2\, w_2\, A_b = \rho_2\, w_2\, \beta A$

Or, $\qquad\qquad\qquad w_2 = w \cdot \frac{\rho_1}{\rho_2} \cdot \frac{1}{\beta}$

At fouling $\qquad\qquad \Pi_2 \equiv \frac{\Delta P_2^*}{\Delta P_2} = \frac{\rho_2^*}{\rho_2} \cdot \left(\frac{\alpha}{\beta}\right)^2 = \frac{1+\tau}{1+\tau\,\delta} \cdot \left(\frac{\alpha}{\beta}\right)^2 = \left(\frac{\alpha}{\beta}\right)^2 \theta_a^{-1}$ $\qquad$ (4-6)

A typical design value for outlet loss fraction is $\xi_2 \approx 0{,}1$.

5. <u>Determination of Air Flow</u>

With (4-3) to (4-6) the pumping power condition (4-2) may now be written in the form

$$\alpha \cong \left[\frac{\frac{\theta_a}{\theta}[1+\xi_1+\xi_2-\kappa]+\alpha\,\kappa\,\delta}{\theta_a\,\theta_b\,\omega^{n-2}+\alpha^{n-1}\,[\theta_a\xi_1+\xi_2/\beta^2]}\right]^{1/(3-n)} \qquad (5\text{-}1)$$

Considering table 3 we note that δ (also part of θ_a, θ_b) is a function of (α, β, ω). Therefore, relative air flowrate α must be solved implicitly. Boundary cases of equation (5-1) are

Forced convection ($\kappa = 0$): $\qquad \alpha \cong \left[\frac{\theta_a}{\theta} \cdot \frac{[1+\xi_1+\xi_2]}{\theta_a\,\theta_b\,\omega^{n-2}+\alpha^{n-1}\,[\theta_a\xi_1+\xi_2/\beta^2]}\right]^{1/(3-n)} \qquad (5\text{-}2)$

Natural draft ($\kappa = 1 + \xi_1 + \xi_2$): $\quad \alpha \cong \left[\delta \cdot \frac{[1+\xi_1+\xi_2]}{\theta_a\,\theta_b\,\omega^{n-2}+\alpha^{n-1}\,[\theta_a\xi_1+\xi_2/\beta^2]}\right]^{1/(2-n)} \qquad (5\text{-}3)$

Note that θ_a and θ_b cancel each other to some extent so that air flow and heat duty are only marginally affected by the temperature effect. There is no eye-catching performance difference of forced or induced fan arrangement when it comes to fouling. For ease of calculation all temperature correction factors in table 3 may be simplified to ≈ 1. If the temperature effect shall be included use equations (3-2) and (3-3) to find the relative temperature difference or alternatively, (3-4) to (3-6). With relative mass flow α identified use equation (3-1) to find the effective heat duty.

If inlet and outlet pressure drop were not covered individually – as is the case in most standards – the form of equations (5-1) to (5-3) would be drastically simplified because $\xi_1 = 0$ and $\xi_2 = 0$. The forced convection case ($\kappa = 0$) would then reduce to

$$\alpha_{00,FD} \cong [\theta\,\theta_b]^{-1/(3-n)} \cdot [\omega]^{(2-n)/(3-n)} \tag{5-2b}$$

and the natural draft case ($\kappa = 1 + \xi_1 + \xi_2$) to

$$\alpha_{00,ND} \cong \left[\frac{\delta}{\theta_a\,\theta_b}\right]^{1/(2-n)} \cdot \omega \tag{5-3b}$$

6. <u>Results Discussion</u>

For discussion of the results a typical multi row water cooler with extruded circular fin tubes has been selected as an example of process cooler. Also, typical design data of a single row condenser system (type ACC) will be used. Basic geometry is listed in table 4.

Design values: $NTU = 1{,}5$; $V = 0{,}4$; $\xi_1 = 15\%$; $\xi_2 = 10\%$; $T_1 = 20°C$ and $\Delta T = 15\,K$, $\tau = 0{,}051168$.

With this large V value we use equation (3-3) ratio of fouled to clean effectiveness for solution to keep rounding errors small especially when fouling layers are wide.

Fin Tube Type	Extruded Multi Row	Single Row
d_t tube transverse diameter	25,4 mm⌀	19 mm (channel 219 mm long)
D external fin diameter	57,2 mm⌀	58 mm (200 mm long wavy sheeting)
s_f fin thickness	0,35 mm	0,35 mm
s_{ext} cladding thickness	0,4 mm	Cladding neglectable
p_t transverse tube pitch	59 mm	58 mm
p_f fin pitch	2,54 mm	2,54 mm
δ_f foot thickness	0,4 mm $=s_{ext}$	0,35 mm
$p_{f0} = p_f - s_f$	2,19 mm	2,19 mm
$d = d_t + 2\,\delta_f$	26,2 mm	19,7 mm
$h_{t0} = p_t - d$	32,8 mm	38,3 mm
$s_{tip} = p_t - D$	1,8 mm	0 mm
$\delta_f^* = \delta_f + \varepsilon\,s_f$	0,4 + ε 0,7 mm	0,35 + ε 0,7 mm
$p_{f0}^* = p_f - s_f - 2\,\delta_{sc}$	2,19 − ε 0,7 mm	2,19 − ε 0,7 mm
$d^* = d_t + 2\,\delta_f^*$	26,2 + ε 0,7 mm	19.7 + ε 0,7 mm
$h_{t0}^* = p_t - d^*$	32,8 − ε 0,7 mm	38,3 − ε 0,7 mm
Heat transfer exponent	$m_k = 0{,}45$	$m_k = 0{,}3$
Bundle pressure drop exponent	$n = 0{,}33$	$n = 0{,}7$

<u>Table 4: Example Data source: (own sket)</u>

Figure (6) shows the results of the above example as function of fouling layer thickness for the case of evenly distributed fouling under forced convection and no contribution (0%) of natural draft. General trends come out as expected - loss of effective flow area ω, heat duty and air flowrate α at rise of fouling. The increase of air temperature difference, Reynolds number and local heat transfer *NTU* – caused by flow acceleration in the gap - does not make up for the total loss of airflow. Whatever local preferences may be - a loss of 8% heat duty ($\approx$ 100% scaling layer thickness, $\approx$ 68% open flow gap area, 82% air flowrate) is generally not acceptable.

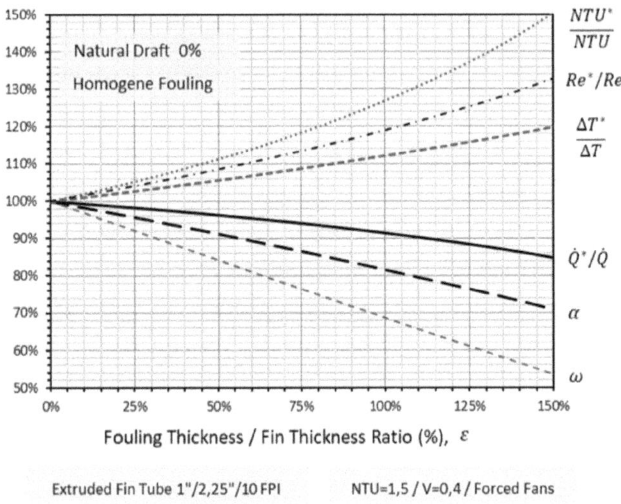

Figure 6: Fouling Performance of Water Cooler – Forced Draft (source: own diagram)

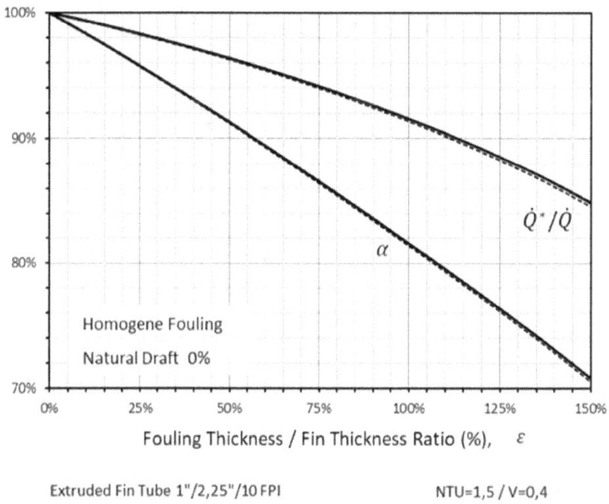

Figure 7: Fouling Performance of Water Cooler – Fan Arrangement (source: own diagram)

Figure 7 shows the results for different fan arrangement using exactly the same data as fig. 6. As already indicated the difference between forced and induced is extremely small with the edge on the side of forced fan arrangement. Consequently, as regards fouling fan arrangement plays no role.

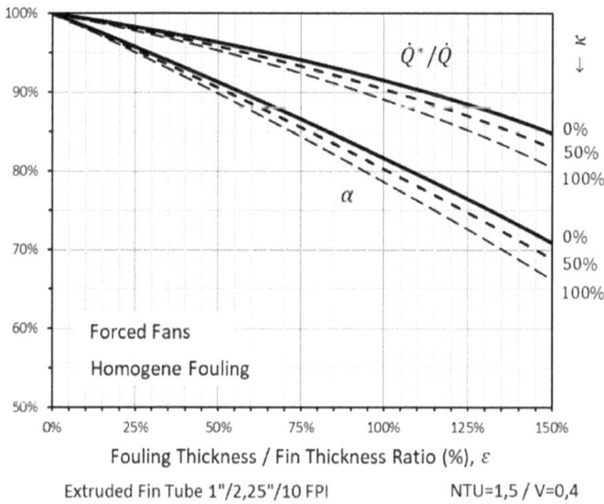

Figure 8: Fouling Performance of Water Cooler – Mixed Convection (source: own diagram)

Figure 8 shows alternative results for different types of convection – forced draft as typical for most process cooling applications, mixed convection (50% natural draft) and pure natural draft as, e.g. in natural draft dry cooling towers. The same water cooler design data as in figures 6 and 7 have been used. Not surprisingly, the sensitivity to fouling rises with increase of natural convection. This calls for more regular cleaning terms in natural draft dry cooling applications than in forced or induced fan arrangements. Note that pure natural draft dry cooling designs are – in tendency – characterized by high *NTU* values which aggravates the fouling problem (cf. fig. 9).

Figure 9 shows fouling heat duty results when *NTU* changes over a typical design range. Two variants – ACC with single row system and, water cooler with extruded fin tubes - have been used. As a general trend – the sensitivity to fouling rises with increase of *NTU* in both types of dry cooling. It is important to note that isothermal ACCs are much more prone to fouling than ordinary water (or process) cooler designs. Again, a high sequence of cleaning must be envisaged in the case of air-cooled condensers. Loss of cooling capacity in ACCs leads to loss of vacuum and generated power which – in turn – may result in dramatic profit consequences. As cleaning is a real issue - fouling characteristics of the planned site should be considered already at design stage. This could well mean to select a different location for the power plant.

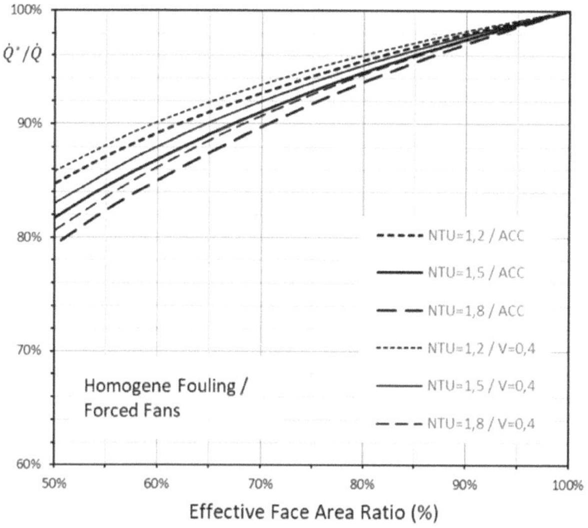

Figure 9: Fouling Performance – Design NTU Effect (source: own diagram)

Figure 10: Large ACC with Single Row System, Egypt (source: Merfort, 2018)

Over the past decades combined cycle power plants with large ACCs have been erected all over the world to meet the rising demand for energy (fig. 10). To reduce internal consumption of these plants it has been proposed to design full natural draft ACCs. Figure 11 deals with the comparison of classical forced draft ACC arrangements to the proposed new ACC type with natural draft as concerns fouling.

To come closer to reality compare forced draft $NTU = 1,5$ with natural draft $NTU = 1,8$ on figure 11. At 80% effective face area we find a loss of heat duty of 5.5% in the case of forced draft, whereas with natural draft the loss is 9%. This substantiates the comments made above.

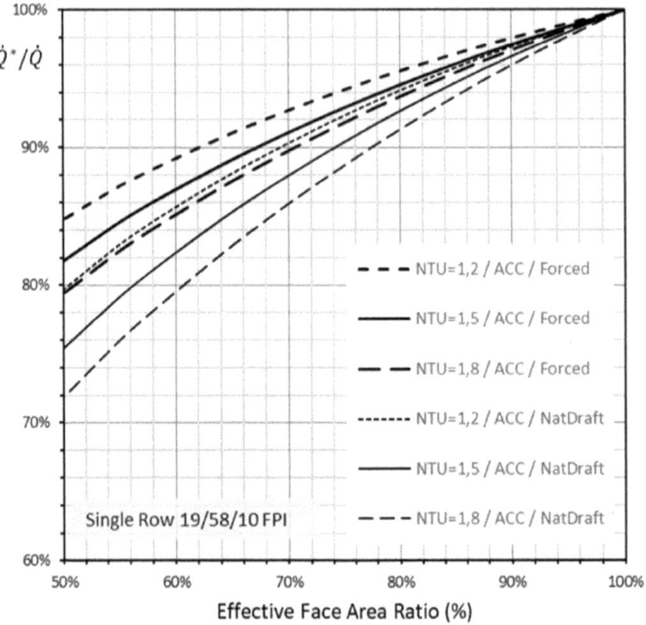

Figure 11: Fouling Performance of ACC – Convection Effect (source: own diagram)

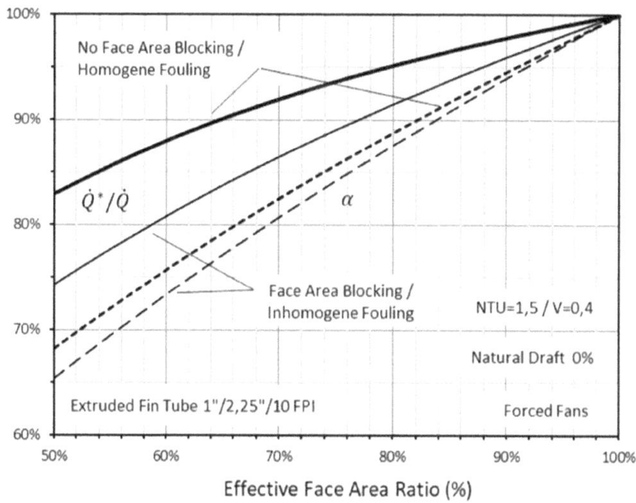

Figure 12: Design Parameter Effect - Partially Blocked (source: own diagram)

The last comparison (figure 12) shows the effect of fouling type – homogeneous fouling (all parts of exchanger active) or, partly blocked exchanger (inhomogeneous fouling). In blocked situation we have assumed that the remaining active part is not fouled. The graphs show the changed air flowrates as well as heat duty as function of the effective face area ratio.

As could be expected - with face area partly blocked the loss of performance is much higher than with homogeneous fouling (for same effective face area). At 90% effective face area the loss of heat duty is doubled in case of partly blocking. It may be concluded that permanent cleaning of a small fraction of the total surface makes good sense – which at best is accomplished with an automatic (or semi-automatic) cleaning device covering the total surface over time.

Glossary

α - air side mass flow ratio (kg/s), fouled/clean

$\beta = A_b/A$ - effective face area factor, un-blocked

$\Gamma_N = \dfrac{(1-V)^2 \cdot NTU}{e^{(1-V) \cdot NTU} - (1-V)}$ - effectiveness auxiliary group (-)

$\Gamma_V = \dfrac{V \cdot (\alpha-1)}{e^{(1-V) \cdot NTU} - (1-V)}$ - effectiveness auxiliary group (-)

$\delta = \Delta T^*/\Delta T$ - air temperature difference ratio, fouled/clean

δ_c - coating thickness (m)

δ_{sc} - thickness of scaling layer (m)

δ_f - fin foot thickness (m)

$\varepsilon = \delta_{sc}/s_f$ - relative fouling (scaling) layer thickness (-)

η_a - average airside viscosity (Pa s)

θ - fan arrangement factor, forced/induced

$\theta_a = \dfrac{(1+\delta\,\tau)}{(1+\tau)}$ - auxiliary temperature ratio

$\theta_b = \dfrac{(1+0,5\,\tau)}{(1+0,5\,\delta\,\tau)}$ - auxiliary temperature ratio

$\vartheta = t_1 - T_1$ - initial temperature difference (K) between process and air inlet

$\kappa = \Delta P_N/\Delta P_B$ - design pressure drop ratio natural draft/bundle - ND factor

$\xi_1 = \Delta P_1/\Delta P_B$ - air inlet pressure drop fraction, design point

$\xi_2 = \Delta P_2/\Delta P_B$ - air outlet pressure drop fraction, design point

$\Pi_B = \Delta P_B^*/\Delta P_B$ - bundle pressure drop ratio, fouled/clean

$\Pi_N = \Delta P_N^*/\Delta P_N$ - natural draft pressure gain ratio, fouled/clean

$\sigma_a = F_a/A$ - surface density (-), heat transfer surface/face area

ρ - air density (kg/m³)

$\Delta\rho = \rho_1 - \rho_2$ - air density difference (kg/m³), inlet (1) - outlet (2)

$\tau = \Delta T/T_1$ - air temperature increase / air inlet temperature, design

$\varphi = F_a/F_i$ -   surface extension ratio: finned to internal tube surface

Φ -   heat exchanger effectiveness (-)

$\omega = \Omega/\Omega^*$ -   relative flow area restriction (-), fouled/clean

$\Omega = A/A_g$ -   face area ratio

A -   total air side face area of exchangers (m²)

A_b -   effective face area (m²), un-blocked part

A_{bg} -   effective minimum gap area (m²), un-blocked part

A_g -   total minimum air side flow gap area of exchangers (m²)

c_A -   air specific heat capacity (J/kg K)

c_P -   mean process specific heat capacity (J/kg K)

d -   maximum transverse core tube diameter (m) including fin root

d_t -   maximum transverse core tube diameter (m)

C -   fin tube pressure drop correlation constant (-)

C_1 -   inlet pressure drop resistance factor (-)

C_2 -   outlet pressure drop resistance factor (-)

D -   maximum transverse over-fin diameter (m)

D_h -   fin tube hydraulic diameter (m)

$f = F_a^*/F_a$ -   surface effectiveness factor at fouling

F_a -   total finned surface (m²)

F_i -   total tube side (internal) surface (m²)

h_a -   air (fin) side heat transfer coefficient (W/m² K)

h_{t0} -   open inter tube height in gap area (m)

H_D -   natural draft height (m)

m -   Reynolds exponent of fin tube heat transfer correlation (-)

m_K -   Reynolds exponent approximation for overall U (-)

NTU -   number of transfer units (-), design

p_t -   transverse tube pitch (m)

p_f -   fin pitch (m)

p_{f0} -   open inter-fin pitch (m)

ΔP_a -   fan pressure drop (Pa)

ΔP_B -   bundle friction pressure drop (Pa)

ΔP_N -   natural draft pressure gain (Pa)

ΔP_1 -   pressure drop at air inlet before bundle (Pa)

ΔP_2 -   pressure drop at air outlet after bundle (Pa)

q -   relative heat duty (-), fouled/clean

$\dot{Q}$ -   heat duty (W)

Re -   Reynolds number at minimum gap area (-)

s_{ext} -   Aluminum cladding thickness (m)

s_f -   fin thickness (m)

s_{tip} -   over fin tip gap (m)

t -   process side temperature (K), at inlet (1) or outlet (2)

$\Delta t_P = t_1 - t_2$ -   process temperature decrease (K), inlet (1) - outlet (2)

T -   air temperature (K), at inlet (1) or outlet (2)

$\Delta T = T_2 - T_1$ -   air side temperature increase (K), outlet (2) - inlet (1)

u -   air velocity in minimum gap (m/s)

U -   overall heat transfer coefficient (W/m² K)

$V = \Delta t_P / \Delta T$ -   capacity flow ratio

$\dot{V}$ -   volume flowrate at fans (m³/s)

w -   air face velocity (m/s)

$\bar{w}_1$ -   air approach speed below inlet (m/s), unspecified flow area

w_2 - air outlet velocity (m/s)

Subscript

1   - air inlet side, forced draft

2   - air outlet side, induced draft

Superscript

*   - with fouling, scaling

Bibliography

[1]   „VGB Guideline Acceptance Test Measurements and Operation Monitoring of Air-Cooled Condensers under Vacuum", VGB R-131 Me, VGB Kraftwerkstechnik GmbH, Essen, 1997.

[2]   „Air-Cooled Steam Condensers, Performance Test Codes", ASME PTC 30.1 - 2007, The American Society of Mechanical Engineers, 2008.

[3]   H.G. Schrey, „Zum Leistungsnachweis luftgekühlter Kondensatoren", Brennstoff-Wärme-Kraft 48 no. 9, paper no. 825, VDI, Düsseldorf, 1996 (in German.)

[4]   H.G. Schrey, „Thermohydraulic Comparison of Fin Tubes", München, GRIN Verlag, Online publication 414394, 2017.